Jean Carlos Feltrin

Thermal comfort in the classroom

Jean Carlos Feltrin

Thermal comfort in the classroom

Case study of the Dante de Mosconi Multipurpose School - Jataí (GO)

ScienciaScripts

Cover image: www.ingimage.com

This book is a translation from the original published under ISBN 978-613-9-63177-3.

Publisher:
Sciencia Scripts
is a trademark of
Dodo Books Indian Ocean Ltd. and OmniScriptum S.R.L publishing group

120 High Road, East Finchley, London, N2 9ED, United Kingdom
Str. Armeneasca 28/1, office 1, Chisinau MD-2012, Republic of Moldova, Europe
Printed at: see last page
ISBN: 978-620-7-77169-1

SUMMARY

ACKNOWLEDGMENTS

I would firstly like to thank my parents Mario Feltrin and Anita Inês Szczepanski Feltrin, my brothers Edilson Feltrin and Carine Inês Feltrin, who have always supported and encouraged me to study, because without them, getting this far would have been impossible. I would also like to thank all my friends and colleagues who have helped whenever possible, even within their own limitations. Special thanks to all the members of the Climatology Laboratory at UFG Regional Jatai, Professor Zilda, Doctor Regina, Master José Ricardo, Academic Elina, among many others, who made me grow as a person and as an academic; to the Dante Mosconi State School and ALL the professionals, who not only became friends, but also taught me to love the teaching profession. Finally, my deepest thanks go to my teachers, especially Dr. Alécio, my advisor, who guided me splendidly, and my academic colleagues in general, who played a part in my career. My thanks to everyone.

SUMMARY

The growth of cities and the occupation of urban land have had a negative impact on thermal comfort. The aim was therefore to analyze and compare the thermal comfort from October to December 2015 in some classrooms at the Dante de Mosconi Polyvalent State School, in the city of Jatai in the Microregion of Southwest Goiás, classified as Zone 6 (bioclimatic strategies) defined as H - Evaporative cooling and thermal mass for cooling, J - Selective ventilation, C- Heavy internal fences according to NBR 15.220-3 (2003). For this purpose, three Thermohygrometers were installed in the classroom in classes A, B and C. The methods used to evaluate thermal comfort were the Fanger method, adopted by ISO 7730 (1987), on a scale of Predicted Mean Vote (PMV) and Predicted Percentage of Dissatisfied (PPD) and the application of a questionnaire. In October, the highest absolute maximum temperatures occurred in rooms A and C, with 34.0°C and 34.9°C respectively. In November, room B had the highest absolute maximum temperature (32.7°C) followed by rooms A (32.6°C) and C (31.4°C). In December, room B had the highest absolute maximum temperature (33.3°C), followed by room C (32.6°C) and room A (32.1°C). The highest PPD indices for the period occurred in October, with A and B having 66.2% and C 59.5%. The other months were also high. In the questionnaire, the majority of respondents in both classrooms said that the classroom was the hottest place in the school, showing that the PPD and the students' opinion are consistent.

Keywords: Thermal comfort, classroom, microclimate.

INTRODUCTION

The disorderly growth of cities and the way in which urban land is occupied, combined with the lack of infrastructure, have caused numerous negative impacts on the environment, including those related to thermal comfort.

Falcao (2009, p.2) emphasizes that "during urbanization, the natural ecosystem was replaced by a completely adverse one, which man organizes according to his survival needs, exercising power over this space".

These transformations end up causing major changes in the characteristics of the environment, which consequently influence climatic elements on a local scale, causing heat islands, places with higher temperatures in urbanized areas than in rural areas.

According to Lombardo (1985, p.77), "the climatic conditions of an extensive urban area with dense construction are totally different from those of the surrounding open spaces, and there may be differences in temperature, wind speed, humidity, air purity, etc.". Generally speaking, the rise in the temperature of the environment has a direct influence on an individual's comfort and their ability to carry out their daily activities, such as studying and working.

The *American Society of Heating, Refrigerating and Air Conditioning Engineers* - ASHRAE, (1997) considers that thermal comfort is defined as a state of mind that reflects satisfaction with the thermal environment that surrounds the person. Therefore, school buildings and classrooms should satisfy students, especially in terms of thermal conditions, since Brazil is a tropical country with high temperatures most of the year.

Ashey (1916) apud Frota (2001, p. 25) in a study comparing thermal comfort with work performance, argues that the first studies on work performance and temperature were carried out in 1916.

> In 1916, under the chairmanship of Winslow, this committee carried out studies and research to determine the influence of thermo-hygrometric conditions on work performance, mainly with a view to the physical work of the laborer, the production interests that arose with the Industrial Revolution and the special situations of war, when troops were deployed to regions with different types of climate. These studies confirmed Herrington's earlier findings:
>
> - For physical labor, increasing the ambient temperature from 20°C to 24°C decreases performance by 15%;

It is therefore necessary to study thermal comfort in schools, in relation to the building standards in their surroundings, as there are not many studies on thermal comfort in school environments, especially in cerrado areas.

Thus, a study on thermal comfort in the classroom was carried out at the Dante de Mosconi Multipurpose School, located in a subtropical climate region with a Cerrado biome, where there are two well-defined seasons: rainy and dry, which can influence the learning performance of these students.

The aim of this work was to collect, analyze and compare the characteristics of the climatological variables of temperature and relative humidity in the classrooms of the Escola Estadual Polivalente Dante de Mosconi in the city of Jatai (GO), seeking to investigate possible differences in the elements of the microclimate with the thermal comfort of the students.

To do this, three Thermohygrometers were installed in three classrooms at the Dante de Mosconi State School to carry out thermal comfort analyses. The structure of the building was analyzed to see if the types of materials used were appropriate to provide thermal comfort, in accordance with NBR 15220-03 (2003). We also sought to analyze whether the type of clothing worn by the students was appropriate and their opinion on the matter.

The Thermal Comfort Index (ICT, USP 2010) and the Arlery, Grisollet and Guilmet Nomogram (1973) were used as measuring models to evaluate thermal comfort.

1. THEORETICAL FRAMEWORK

1.1 Thermal comfort

As a result of urbanization, Mendonça and Monteiro (2003, p. 19) describe the concept of Urban Climate as "(...) a system that encompasses the climate of a given terrestrial space and its urbanization."

Understanding the phenomenon of urbanization as an element to be considered in climatology studies, and considering that the classrooms are inserted in the urban environment, the Monterian theory was indispensable for studies in which the urban environment was the stage for the unfolding of the research.

As pointed out by Christofoletti (2012, p. 714), concepts such as "Urban Climate", as elucidated by Monteiro and Mendonça (2003), have opened up more space for studies of "Thermal Comfort", "Thermal Sensation", "Thermoregulation", as addressed by Frota (2003), and "Thermal Comfort Indices".

For Frota and Schiffer (1998), the term thermal comfort is defined by the human organism's need for thermal exchange between its metabolism and the environment in which it lives, without resorting to thermoregulation mechanisms. This process of adaptation to the environment is therefore called thermoregulation. Ruas (1999, p. 14) explains this as follows:

> (...) Man is a homothermic being, i.e. he can maintain a relatively constant internal body temperature within certain limits, regardless of the ambient temperature. Therefore, there must be permanent and immediate elimination of the excess heat produced so that the body's temperature can be kept constant. Body temperature is controlled by a system called the thermoregulatory system, which controls, through vasodilation and vasoconstriction, the amount of blood circulating on the surface of the body, allowing, respectively, a greater or lesser exchange of heat with the environment. The thermoregulatory system also acts on the sweat glands, increasing or decreasing sweat production depending on the body's need to lose heat through sweat evaporation."

Despite the terms used, they both deal with the relationship between the environment and the biological balance with it, which is why, according to Lamberts et al. (1997), thermal comfort is defined as "a state of mind that reflects satisfaction with the thermal environment that surrounds the person". Thus, it is understood that the thermal sensation is relative from one individual to another and also depends on each person's metabolism.

The ASHRAE standard (American Society of Heating Refrigerating and Air Conditioning Engineers, 1981) describes that a room should have thermal conditions that at least 80% of occupants feel thermally satisfied in.

In this way, the study of Thermal Comfort demonstrates the importance of providing a thermally pleasant environment in which to carry out activities, such as teaching and learning activities.

1.2 Climatic Factors and Thermal Comfort

The factors to be considered when analyzing thermal comfort are diverse and very complex if added together, which is why most studies evaluate the main ones such as rainfall, air temperature, relative humidity, wind, geographical aspects of the region, clothing, solar radiation and structure.

1.2.1 Rainfall

The condensation of water vapor in the form of rain originates largely from rising masses of humid air, which are cooled rapidly by contact with colder air masses. As precipitation occurs, temperatures tend to drop and relative humidity rises. According to Lombardo (1985) researching the climate in the city of São Paulo, an accumulated rainfall in 24 hours of around 18 mm causes a sharp drop in local temperatures.

According to Silva (2012, p. 38) in his study on precipitation and thermal comfort, he describes that the results clearly demonstrated the positive effect of precipitation on temperature in the city of Manaus, acting as a mitigator of local thermal discomfort.

In this way, precipitation becomes a very important element for analyzing thermal comfort. Knowing when it falls more or less frequently is of the utmost importance in a study area.

1.2.2 Air temperature

On our planet, there are different types of relief with different altitudes that receive more or less solar radiation, resulting in different temperatures. The positioning of the planet in relation to the sun in its process of translation, associated with different solar radiation absorption coefficients, different types of soil and water disposition, lead to an unequal distribution of solar energy and will also cause the temperature to change, leading to the seasons, as described by Dias (2009, p. 49).

Most of Brazil is situated below the equator (Southern Hemisphere), but there are no mountain ranges or higher altitudes, and most of it is intertropical, with less severe winters (temperatures do not go below zero in most of the country), while summers tend to be hotter and more humid.

Air temperature, despite being influenced by a variety of factors, is one of the most important factors when it comes to thermal comfort. An increase or decrease in temperature directly affects the human organism, which responds to stimuli in order to get back into balance, producing more or less energy through metabolism.

According to Rocha (2012), the city of Jatai has an average annual temperature of 22.0°C, which characterizes it as a warm climate.

From the point of view of thermal comfort, this temperature would not be far from the ideal for the balance between the human organism and the environment, but other factors besides temperature must be considered in order to know whether or not the environment is thermally comfortable for the individuals who occupy it. Nor should you only work with average temperatures, as the most important thing for thermal comfort studies is extreme temperatures.

1.2.3 Average radiant temperature

Radiant temperature is defined by NBR 15220-3 (2003) as "the uniform temperature of an imaginary environment where radiant heat transfer occurs between the human body and a non-uniform real environment".

In a classroom, the mean radiant temperature (MRT) according to Nogueira et al (2005) can be obtained from the dry bulb temperatures using the formula:

$$TRM = \frac{\sum_{i=1}^{n}(T_i . A_i}{\sum_{i=1}^{n}(A)}$$

where:

TRM: Average radiant temperature

T_i = temperature of the room surfaces, in °C;

A_i = surface area in m^2

Although the author explains how the formula is applied, he explains that in his work it was calculated using Analysis 1.5 software.

The average radiant temperature is fundamental for thermal comfort, as it deals with the energy that a body can emit into the environment and vice versa, making its analysis extremely important.

1.2.4 Relative humidity

Relative humidity is the amount of water particles that have evaporated and are present in the air. In places where there is a lot of solar radiation and a lot of water, the humidity tends to be higher.

Relative humidity also affects people's thermal comfort, according to Rodrigues (2014, p. 3)

> It should be noted that a reduction in relative humidity to below 30% is considered a risk to the integrity of the airways, hindering the internal homeostasis of the respiratory system. In the rainy months, the factors that can contribute to an increase in respiratory diseases, especially allergic diseases, are: high relative humidity, more time spent indoors, less ventilation and exposure of home spaces to the sun and consequent growth of mold and fungi.

According to the Emergency Management Center of the City of Sao Paulo (CGE, 2015), a period in which the relative humidity of the air is low can cause complications.

- Allergic and respiratory complications due to dry mucous membranes;
- Bleeding from the nose;
- Skin dryness;
- Eye irritation;
- Static electricity in people and electronic equipment;
- Increased potential for fires in pastures and forests.

In this way, the Center for Meteorological and Climatic Research Applied to Agriculture (CEPAGRI, 2015) also recommends some precautions in order to try to avoid as few impacts as possible on people's health with low humidity.

Between 20 and 30% - State of Attention

- Avoid outdoor exercise between 11 am and 3 pm
- Humidify the environment using vaporizers, wet towels, containers of water, watering gardens, etc.
- Whenever possible, stay in places protected from the sun, in vegetated areas, etc.
- Drink plenty of water

Between 12 and 20% - State of Alert

- Observe the recommendations of the state of attention
- Suppress physical exercise and outdoor work between 10 a.m. and 4 p.m.
- Avoid crowding indoors
- Use saline solution for eyes and nostrils

Below 12% - State of emergency

- Observe the recommendations for alertness and alertness.
- Order the interruption of any outdoor activity between 10 am and 4 pm, such as physical education classes, garbage collection, mail delivery, etc.
- Order the suspension of activities that require crowds of people indoors, such as classes, cinemas, etc. between 10 a.m. and 4 p.m.
- In the afternoons, keep indoor areas humid, especially children's rooms, hospitals, etc.

1.1.5 Winds

Wind is a factor that, together with temperature and humidity, interferes with people's thermal comfort. According to Dias (2009), the different atmospheric pressures on our planet caused by the variation in the incidence of solar radiation and the varied distribution of continental masses cause places of low and high atmospheric pressure, causing air masses to constantly move around the globe.

According to Frota (2001) the difference in pressure exerted on the building can be caused by the action of the wind. The wind, which moves parallel to the ground in a lamellar motion, when it encounters an obstacle (building), suffers a deviation of its fillets and, when it passes the obstacle, tends to resume the lamellar regime.

In this way, the local climate and the geographical orientation of a building have a direct influence on whether or not the thermal load is dissipated, since all buildings generate thermal load through machinery, equipment, lighting, people and physical activity.

Knowing this, in order to dissipate the thermal load, natural air circulation becomes necessary within a built environment, as well as contributing to the renewal of indoor air (removal of indoor air pollutants), which, depending on the occupancy profile of the environment, can affect the productivity of the occupants as well as being harmful to health.

According to Bittencourt and Cândido (2008) apud Pereira et al (2012, p. 3), natural ventilation regulates the internal climate of a building. This occurs through an exchange of air controlled by openings at the top and bottom of the structure, the so-called chimney effect.

The chimney effect works through the temperature difference between the outside air and the air inside the built environment and the pressure differences caused by the action of the wind.

At the top of the built-up area is the high-pressure zone, where openings are created so that this warmer air is "pushed" out by the colder air that will enter through the lower opening (low-pressure point). The warmer air is lighter because its particles are further apart and more agitated, so it tends to rise. The colder air is heavier and its particles are closer together and less agitated, so it tends to descend, thus creating the chimney effect shown in Figure 1.

Figure 1 - Chimney effect: heavy air in and light air out

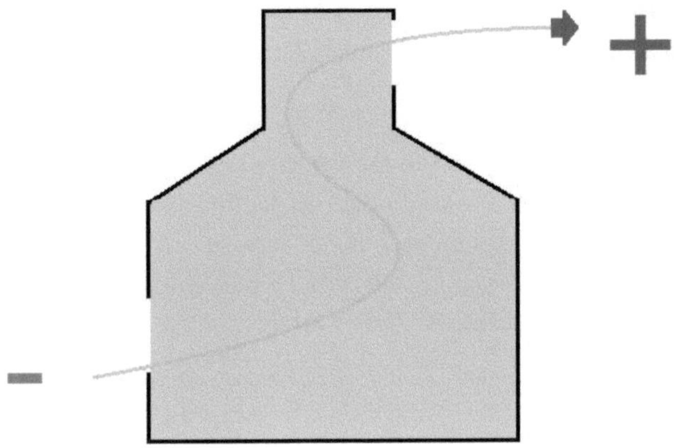

Author: FELTRIN, 2016.

Knowing this, we can obtain an environment with more suitable thermal conditions by using only load exchange, as emphasized by Pereira et al (2012, p. 2)

> A clean, cheap and sustainable alternative for improving the above scenario is the use of natural ventilation to promote thermal comfort in the workplace and to reduce fixed electricity costs. Natural ventilation is the phenomenon of air movement inside buildings without the induction of any mechanical or artificial system, generated by a difference in air pressure, which can occur due to the action of the wind or a difference in air density due to a difference in temperature. In both processes, there must be inlet and outlet openings so that the air can flow through the building.

The orientation of the buildings is also fundamental for the load exchange process to take place, since the wind and the position of the sun change little according to the seasons, so constructing a building with strategic parts facing more to the north, south, east or west can help to mitigate the concentration of loads.

1.1.6 Geographical aspects and the influence of continentality

There are geographical aspects such as longitude and latitude, which are imaginary lines created to help with the distribution of hours, and hemispheres. In the case of longitudes, they are measured in relation to the Greenwich Meridian. Every 15° is one hour in the time zone, giving a total of 24 hours, i.e. 360°, one complete turn and therefore one day.

Latitude is measured from the equator, imagining a line that circles the surface of the Earth, halfway between the two poles, i.e. the middle between the south and north poles. Latitude is measured from 0° to 90° and is said to be South if it is below the equator and North if it is above.

According to Ayoade (2003), the main control over the amount of sunlight a given place receives is a function of its latitude. This is because the variation in the angle of incidence of the sun's rays and the length of the day in any given place are determined by its latitudinal location.

As Frota (2001) also tells us.

> If the imaginary axis joining the poles were perpendicular to the plane of the ecliptic, which is the Earth's plane of translation around the Sun, every point on its surface would see the Sun in the same position throughout the year. But since this axis is inclined at approximately 23 1/2° to the normal, [...] the Sun will apparently travel through a region of the sky corresponding, on Earth, to that between the tropics of Cancer and Capricorn, with a duration of six months in each direction.

This process gives rise to the four seasons. The latitude of a region, combined with the time of year, will determine the amount of sunlight.

Again, Ayoade (2009) describes that

Seasonal variations in temperature are mainly the result of seasonal variations in the amount of sunlight received anywhere on the globe. Temperatures are higher in summer, when the amount of sunlight is greatest, and lower in winter, when the amount of sunlight received is lowest.

With this in mind, continentality is a concept used in geography that deals with the climate of places that are far from the sea. Due to the remoteness of these locations, the air temperature and humidity tend to be different from those on the coast because of the sea.

Water absorbs more heat, causing its temperature to rise by the same amount as the soil. Therefore, the closer you are to large bodies of liquid, the smaller the temperature variation (Maritimality) and the further you are from large bodies of liquid, the greater the temperature variation (Continentality). Therefore, continental climates are characterized by large temperature variations.

This process can be explained by the fact that in the winter, the central region of Brazil, where Jatai is located, has higher temperature ranges and lower relative humidity, which can drop below 15%.

1.3 Influence of clothing

According to Frota (2001) clothing represents a barrier to heat exchange or water loss through perspiration, and by keeping a layer of air still, it hinders exchange through convection and radiation. The use of clothing varies according to the climate the person is in and their biological thermal comfort. There are clothes that can minimize the amount of moisture to be transpired, or act as a thermal insulator by keeping a layer of warmer or less warm air close to the body, and also reduce the body's sensitivity to variations in temperature and air speed. Its thermal resistance depends on the type of fabric.

According to Barros (2001), around 20% of this energy is transformed into work potential and the other 80% is transformed into heat, which must be dissipated in order for the body to remain in balance. Therefore, there are garments made of materials that retain little heat and others whose material retains more heat, and you just have to choose which is more appropriate for the environment you are in.

The study site wears a uniform consisting of jeans and a short-sleeved cotton T-shirt. Hats, hoodies or any other garment worn on the head are not allowed, but there is no requirement as to the footwear to be worn.

As a basis for the thermal insulation of the clothing and the individual's metabolic rate, we used the constant measurements made by ASHRAE (1997).

1.4 Solar Radiation and Building Standards

When the sun shines on a building, there is always a heat gain which will be a function of the intensity of the incident radiation and the thermal characteristics of the building's parameters, which vary according to the material used.

Therefore, a building can absorb more or less radiation according to the direction in which it was built. Buildings facing north tend to receive more solar radiation if they are located in the southern hemisphere, which is the case for most of Brazil. Using this information, buildings can be planned so that they receive sunlight during the cooler parts of the day and less sunlight during the hotter periods. In this way, the building will receive less solar radiation and therefore tend to be cooler.

With this in mind, NBR 15220-3 has a reference to thermal comfort for thermal conditioning in buildings by Brazilian bioclimatic zoning and construction guidelines for the entire national territory. In the region of Jatai (GO), NBR 15220-3 establishes bioclimatic zone 6 (Z6), Figure 2.

Figure 2 - Brazilian Bioclimatic Zones (NBR15220-3)

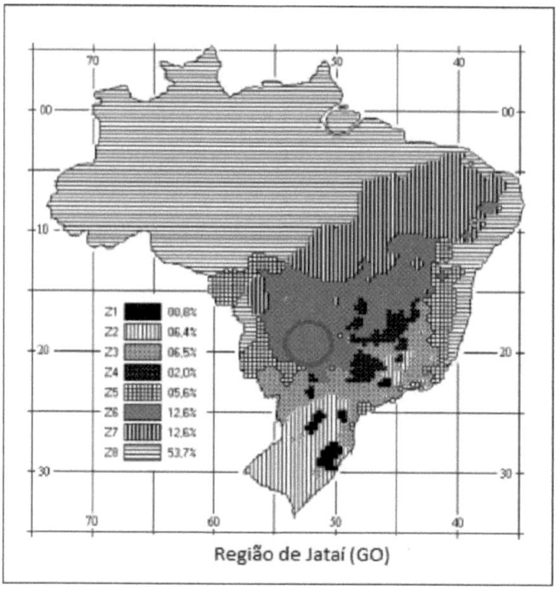

Região de Jataí (GO)

Source: NBR15220-3, 2003

According to Standard NBR15220-3, Z6 has a "Thermal Conditioning Strategy" classified as H, J, C.

Table 1 - NBR15220-3 recommendations

Summer	H - Evaporative cooling and thermal mass for cooling
	J - Selective ventilation (during hot periods when the internal temperature is higher than the external temperature)
Winter	C- Heavy internal seals (thermal inertia).

Source: NBR15220-3

Organizer: FELTRIN, 2016

As for ventilation, NBR15220-3 establishes that the ventilation opening should be "Medium", i.e. 15 to 25% of the floor area, and that the shading of the openings should be "Shaded".

Wall seals should be "Heavy", with a thermal lag (φ) of ≥6.5 hours and the roof "Light and Insulated" with φ ≤ 3.3. Table 2. Foot wheel

Table 2 - Ventilation openings and wall seals suggested for Z6

Example figure	**Description**	**φ hours**
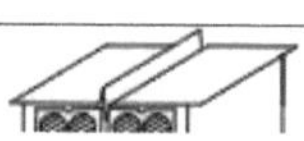	Double wall of 6-hole circular	6,5

	bricks, laid to the smallest dimension Brick dimensions: 10.0x15.0x20.0 cm Laying mortar thickness: 1.0 cm Plastering mortar thickness: 2.5 cm Total wall thickness: 26.0 cm	
	Fiber cement tile roof with concrete lining tile thickness: 0.7 cm concrete thickness: 3.0 cm	2,6

Source: NBR15220-3

Organizer: FELTRIN, 2016

However, it is common to find schools whose floor plans are the same as those of other institutions in different parts of the state and country, because in order to contain costs, construction is standardized and does not always follow the recommendations of NBRl 5220-3, since it is possible to save material by using a single wall instead of a double wall as is recommended, for example.

The positioning of the structure in relation to the cardinal points is also often not possible or not taken into account, making the structure inefficient from the point of view of natural charge exchange, as already mentioned.

1.5 Measuring Thermal Comfort

One way to analyze Thermal Comfort is to apply its variables in indices, so we chose the Thermal Comfort Index (ICT, USP 2010) and the Nomogram of Arlery, Grisollet and Guilmet (1973), whose results are representative values of Thermal Sensations, which point to a comfortable or uncomfortable environment.

The Thermal Comfort Index (TCI) relates dry bulb temperature in degrees Celsius (T) and relative humidity in percentages (RH) as follows: TCI = 0.55 (1-0.01UR) (T - 14.5).

The ICTs are classified as: <10 - Feeling very cold; from 10.1 to 14.9 - Feeling cold; from 15 to 19.9 - No discomfort (ideal); from 20 to 24.9 - Great discomfort in the heat; > 25 - Maximum discomfort in the heat.

Another index used, developed by Arlery, Grissollet and Guilmet in 1973 for tropical climates, is called the "Arlery, Grissollet and Guilmet Nomogram", which relates temperature in degrees Celsius to relative humidity in percentages, generating areas on the graph that define thermal sensations in six classes.

The Analysiss 1.5 software, made available to the public by the Federal University of Santa

Catarina (UFSC), was used to calculate the indices based on Fanger (1970). In order to calculate the indices, the software needs the following values: dry bulb temperature, wind speed, relative humidity, average radiant temperature, clothing resistance and metabolic rate.

2. MATERIAL AND METHODS

2.1 Characterization of the Study Area

For this study, we chose the Dante de Mosconi Polyvalent State School, located in the city of Jatai (Map 1), in the South-West Microregion of Goiás, classified as Zone 6 with bioclimatic strategies defined as HJC, according to the classification of NBR 15.220-3 (Table 1), with a territorial area of 7,174 km2 and a population of 88,006 inhabitants (IBGE, 2010).

Map 1 - School Location

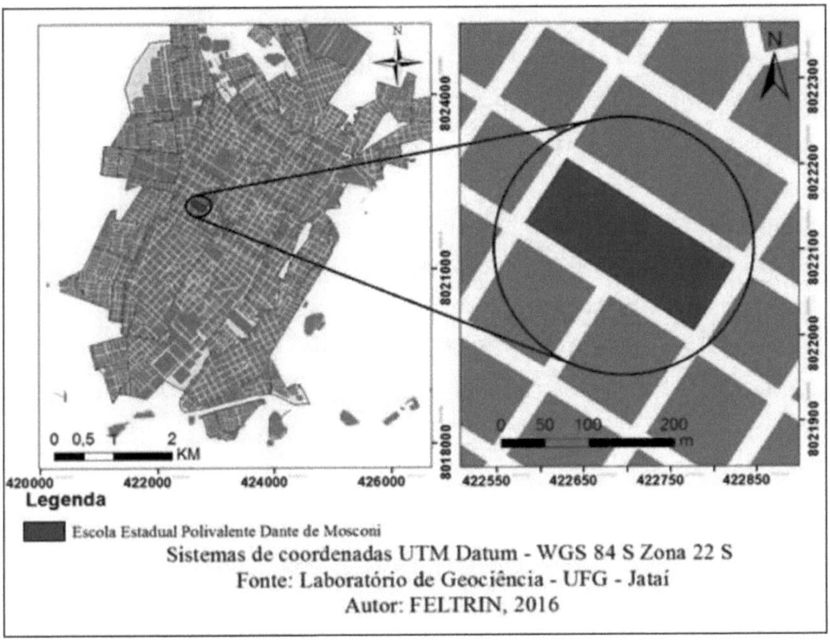

The municipality of Jatai has a tropical mesothermal climate, with two well-defined seasons: rainy and dry. The rainy season begins in October and lasts until April, when the dry season begins and lasts until September. According to Mariano and Scopel (2000), the Jatai region has a total rainfall of between 1600 and 1700mm, but as already mentioned, rainfall is distributed throughout the rainy season.

Due to anthropogenic actions, it is necessary to study thermal comfort in urban areas, but there are not many studies on thermal comfort in the classroom, given that the study site is in a subtropical climate region with a Cerrado biome, far from large urban centers. Studying thermal comfort in the classroom in the Cerrado can provide a broader understanding of how students are influenced and better plan education as a whole, structure, timetable, vacation period, etc.

The school occupies an area of $19,994.00m^2$, of which approximately 6,667.42 m2 is built area. There is a large courtyard where students can sit during breaks and the area is well wooded.

Its physical structure is reasonably well preserved, as it has received some interventions from time to time, such as renovations and extensions, which have left it looking good and in good condition for use.

Part of the building has ten classrooms measuring approximately 60m2 each, a headmaster's office, deputy headmaster's office, pedagogical coordinator's office, teachers' room, boys' and girls' toilets with twelve (12) rooms in each block, one (1) changing room with electric showers for boys and girls, a large library, a room for the school secretary, a kitchen, a storeroom, a covered patio for daily socializing and artistic and cultural activities. The unit also has a sports court, where practical physical education classes are held.

The Escola Estadual Polivalente de 1° Grau was created on April 22, 1977. The school's name was changed by Law No. 8,432 of April 20, 1978, published in the Official Gazette of the State of Goiás, No. 12,998 of April 27, 1978, page 2, and renamed the Dante Mosconi Polyvalent School. The school was named after civil engineer Dr. Dante Mosconi, who was born in Italy on November 1, 1892. After living in various regions of Brazil, he came to Jatai, where he dedicated himself body and soul to teaching in the city of Jatai.

2.2 Material

To carry out the study, we used three Thermohygrometer devices (Data logger model HT-500) (Photo 1) and three shelters to protect the devices from falls, insects, solar radiation, wind, etc. arranged in three classrooms of the school's ten classes. A notebook computer was used to download the data, and Excel software was used to create the spreadsheet database.

Arcgis 10.1®, licensed to the Geoinformation Laboratory of UFG/Regional Jatai, and Google Earth Pro, with images from 2014, were used to survey the built-up area and the location map.

Photo 1 - A - TlierimoliigmimetiO; B - Shelter

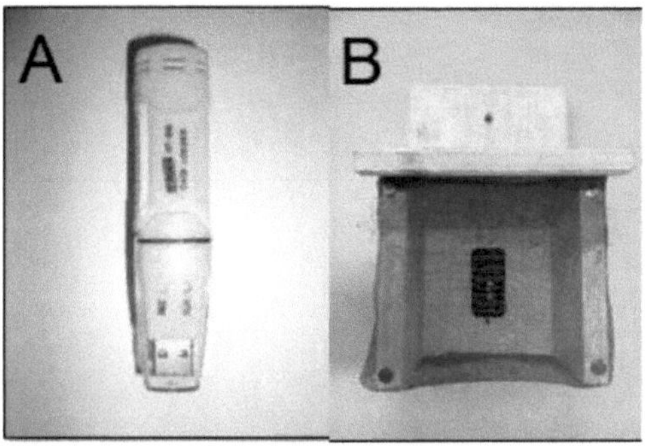

Author: FELTRIN, December 18, 2015

2.3 Methodological Procedures

Before starting data collection, the equipment was tested and the height of the equipment chosen.

2.3.1 Equipment testing

The devices were labeled so that they were always used in the same places and to avoid any confusion when saving the information. The three devices were also placed next to each other in a cabinet to check that they were calibrated. The results showed that the device named B was 0.7°C lower than A, and C was 0.5°C higher than A, so Datalogger A became the reference for calibration. After the analysis, it was decided that the variation values would be corrected by deducting these differences from the spreadsheet.

During the test period, with the equipment connected and under the same conditions, they showed that it takes 10 to 15 minutes for the measurements to stabilize, a time that was determined when the equipment was installed before school.

2.3.2 Defining the installation height of appliances

After adjusting the temperature variation, the height at which the equipment would be placed for data collection was analyzed. Three shelters with three devices were placed inside the classroom at different heights. The first device was placed at a low height (0.5m), the second at a medium height (1.5m) and the third at a high height (2.2m), Photo 2.

Photo 2 - Testing the height of the equipment

Author: FELTRIN, November 10, 2015.

Graph 1 shows that the temperature was higher at a height of 1.5m. It was therefore decided that the height to be followed would be 1.5m, the same height used by Rocha, Mariano and Feltrin (2014).

Graph 1- Temperatures at different heights in the classroom

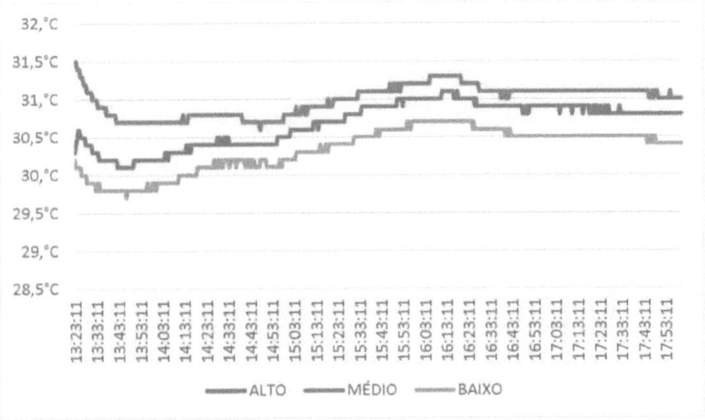

Author: FELTRIN, 2015.

2.4 Choice of Analysis Rooms and Equipment Installation

The school has ten classrooms, from which three classes were chosen to install the Thermohygrometers, each with a shelter. The classrooms chosen were: C, A and B, Figure 3.

The devices were positioned at the back of the classrooms at a height of 1.5m from the wall and in the middle of the wall, so as not to interfere with the progress of the lessons, the students' view or their movement (Photo 3).

Photo 3 - Location of the Thermohygrometer in the classroom

Author: FELTRIN, November 10, 2015.

We also chose a room at each end of the block (C and A) and one in the center (B), looking for rooms of the same size that had both ceiling fans working. We also chose rooms with different types of ceilings, since the type of material can affect the room's temperature. Therefore, of the three rooms chosen, only room C had a slab ceiling and the other two had wooden ceilings. Room C is on the left-hand side of the block, facing north. It has gray-tinted windows and a slab ceiling, and there is a ramp outside which shades the windows during part of the day.

Figure 3 - School block and the layout of the classrooms studied

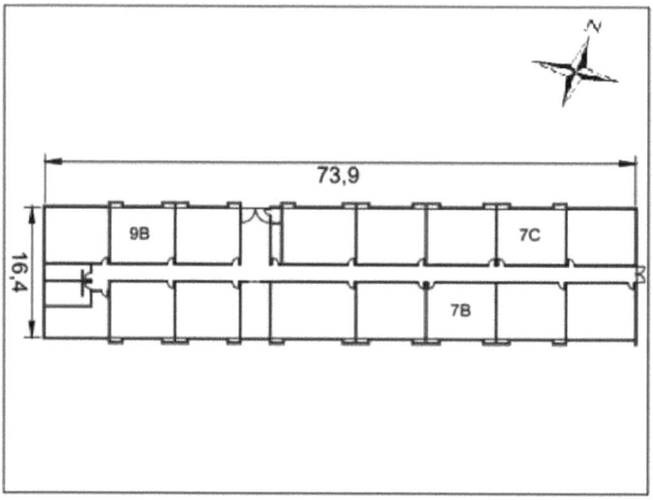

Author: FELTRIN, 2016

Rooms A and C are facing more to the northeast, and room B is facing more to the southeast. As a result, the rooms that tend to receive the most solar radiation are A and C, as shown in Figure 3.

The collection of temperature and relative humidity data began on October 7 until December 18, 2015. To begin the collection, the devices were installed 15 minutes before the start of class (12:45 pm) to stabilize and removed at 6:00 pm to download the data into a notebook in spreadsheet format and then organized into a database to facilitate visualizations of possible eventualities.

2.5 Fanger method

The Fanger evaluation method, proposed by the IPT and adopted by ISO 7730 (1987), aims to determine the degree of thermal comfort or discomfort in an environment, following a scale ranging from +3 to -3 (Table 3).

Table 3 - Degree of thermal comfort or discomfort in an environment proposed by the IPT and adopted by ISO 7730

+ 3 very hot	+ 2 hot	+ 1 slight sensation of heat	0 comfortable	- 1 slight sensation of cold	- 2 cold	- 3 very cold

Author: FELTRIN, 2016

This method uses parameters such as air temperature (T°C), relative humidity (RH%), relative air speed (m/s), average ambient radiant temperature (°C), clothing worn by the individual (clo) and the individual's metabolism (w/m^2) depending on the activity carried out, as established by ASHARAE.

Fanger's equation can be defined as follows:

$$PMV = (0{,}303e^{-0{,}036M} + 0{,}028)\,\{(M - W) - 3{,}05x10^{-3}x\,[5733 - 6{,}99(M - W) - \rho_a] - 0{,}42x[(M - W) - 58{,}15] - 1{,}7x10^{-5}M(5867 - \rho_a) - 0{,}0014M(34 - t_a) - 3{,}96x10^{-8}f_{cl}x[(t_{cl} + 273)^4 - (t_r + 273)^4] - f_{cl}h_{el}(t_{cl} - t_a)\}$$

Where PMV - Estimated mean vote M - Metabolic rate (W/m2), W - Mechanical work (W/m2) - null for most activities, p_α - Partial pressure of water vapor t_α - Air temperature (°C) f_{ci} = - Ratio between the surface area of the clothed body and the surface area of the naked body t_{cZ} - Surface temperature of clothing (°C) t_r - Average radiant temperature (°C) hc - Convective heat transfer coefficient (W/m2°C).

Analysis 1.5 software was used to carry out the PMV and PPD calculations. The software is very simple and easy to use, all you have to do is enter the values and the software performs all the mathematical calculations, displaying the PMV and PPD values (Figure 4).

Figure 4 - Analysis 1.5 software home page

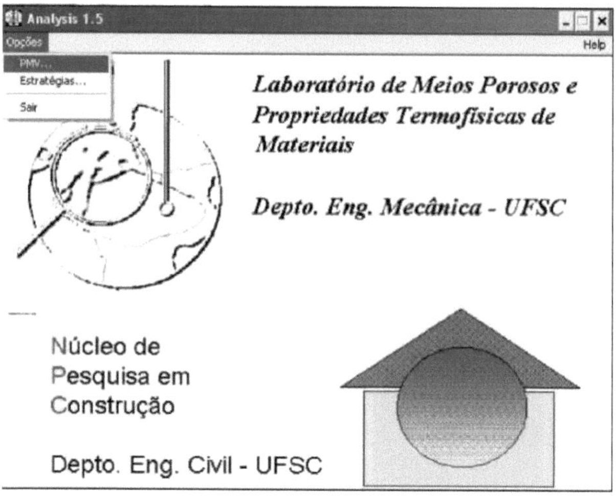

Source: LabEEE, 2016

After clicking on options, a new window opens in which you can choose the clothes and activity that the individuals are doing. In the case of the students, sitting down and relaxing was the option chosen and the clothes were those that make up the school uniform (Figure 5).

Figure 5- Choice of clothing and activity carried out

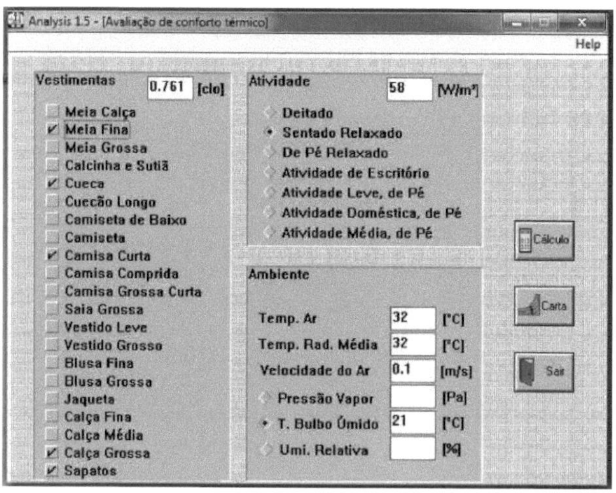

Source: LabEEE, 2016

After choosing the clothing options, the activity carried out, and the temperature values, average radiant temperature and wet bulb temperature, click on the calculate option. The software will open a new window with the PMV and PPD values (Figure 6).

Figure 6 PMV and PPD results performed by the software

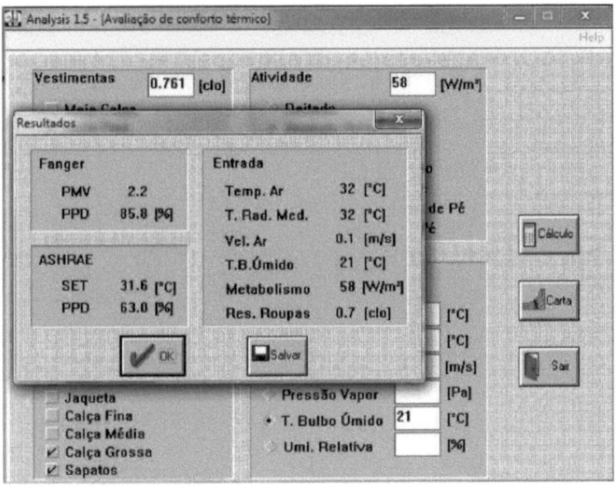

Source: LabEEE, 2016

2.6 Preparation and application of questionnaires

The questionnaires were drawn up taking into account the reality of the classrooms and, based on these, questions were asked about the students' perception of reality, considering thermal comfort.

There were 14 identical questions for the three rooms, applied on the same day and at the same time, considering the same weather situation. To facilitate the answers and their quantification, most of the questions were objective. These questions asked about the type of clothing they were wearing, how they would rate the lighting, ventilation and temperature, for example (Appendix A).

For the non-objective questions, we tried to ask more open-ended questions where the students could give broad and varied answers, such as "What clothes do you find most comfortable to wear in the classroom according to thermal comfort?".

Classroom C has some different characteristics, which is why two more questions were asked in this classroom, making a total of 16 questions. The questionnaires were administered on January 22, 2016, when the students were returning to school to make up for classes missed during a strike in 2015.

The questionnaires were based on models drawn up by Xavier (1999, p.133) and Grandi (2006, p.133) and later adapted to the objectives and reality of the site.

3. RESULTS AND DISCUSSION

2.7 Infrastructure Analysis

Room A measures 7x8.5m, with a total area of 59.5m^2 There is a single entrance on the left, which gives access to the main corridor of the school block. The room has two windows, one with larger dimensions on the right side measuring 5.3x1.9m, giving a total area of 10.1m2, and another window on the left measuring 0.7x5.9m, giving a total area of 4.13m2 (Figure 6).

Figure 7 - Measurement of walls and windows in room A

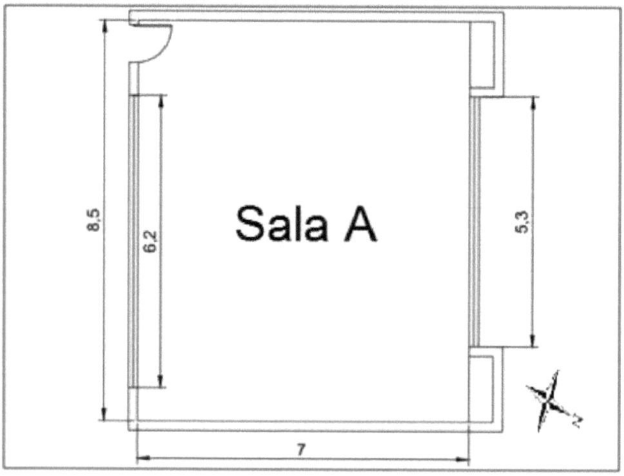

Author: FELTRIN, 2016

If we add up the two window areas, we get 14.23m2, while the floor area is 59.5m2. The window area then corresponds to 23.9% of the floor area. According to NBR 15220-3, this value should correspond to 15 to 25%, so the structure has the measurements suggested by the regulations.

The walls of the classrooms are 0.2m (20cm) thick on all sides, which is not in line with what is suggested by NBR 15220-3 (2003), which recommends

Double wall of 6-hole circular bricks, laid to the smallest dimension; brick dimensions: 10.0x15.0x20.0 cm thickness of laying mortar: 1.0 cm thickness of plastering mortar: 2.5 cm total wall thickness: 26.0 cm (NBR 15220-3)

The ceiling in the living room does not meet the standards "Fiber-cement tile roof with concrete lining tile thickness: 0.7 cm concrete thickness: 3.0 cm". The ceiling is made of wood and not concrete, *while the* roof complies with the standards, being made of fiber cement (NBR 15220-3, 2003), Photo 4.

Photo 4 - Classroom ceiling A

Author: FELTRIN, November 10, 2015.

Classroom B also measures 7x8.5m, totaling 59.5m^2 with a single entrance on the right-hand side giving access to the main corridor of the school block. The room also has two windows with the same dimensions: a window on the left measuring 5.3x1.9m with a total area of 10.1m2, and another window on the right measuring 0.7x5.9m with a total area of 4.13m2, with the same measurements as room A and therefore the same area, as suggested by NBR 15220-3.

The walls of the classrooms are 0.2m (20cm) thick on all sides, not complying with the standards suggested by NBR 15220-3, mentioned above. The ceiling of the classroom does not meet the standards, as it is made of wood and the roof is fiber cement, which complies with NBR 15220-3 (Figure 7).

Figure 8 - Measurement of walls and windows in room B

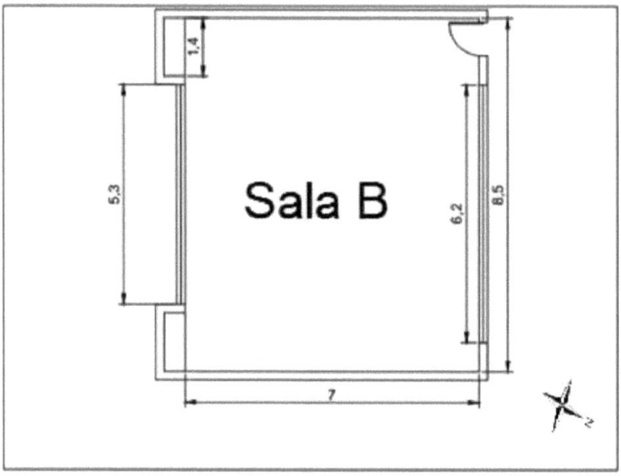

Author: FELTRIN, 2016

Classroom C measures 7x8.15m, totaling 57.1m^2 and is smaller than the other two classrooms, with a single entrance located on the right-hand side, which also gives access to the main corridor of the school block. The room has two windows on the right with dimensions of 3.4x1.7m, totaling 5.78m2 of area of each window, if added together we have 11.56m2, corresponding to 20.3% of the floor area, being a smaller area than the other rooms, but meeting the specifications of the NBR 15220-3 standard, which is 15 to 25% (Figure 8).

The walls of classroom C are 0.2m (20cm) thick, also proving to be school block standard. It therefore does not comply with the aforementioned NBR 15220-3 standard.

The ceiling of the room meets the standards, being made of concrete and the roof is fiber cement in accordance with NBR 15220-3. The windows of the room are painted gray, blocking the entry of natural light, and the area of the room is smaller, these being the main differentials of the room in relation to the other two (Photo 5).

Photo 5 - Concrete ceiling and painted glass in room C

Author: FELTRIN, November 10, 2015.

Figure 9 - Measurement of walls and windows in room C

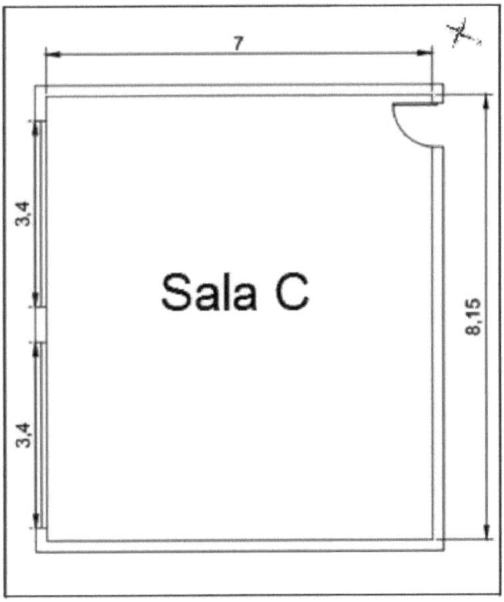

Author: FELTRIN, 2016

2.8 Thermo-hygrometric analysis of the classroom

2.8.1 Absolute maximum and minimum temperatures

Rooms B and C had the highest absolute temperatures in October, both with 34.9°C. Room A had the lowest absolute maximum temperature at 34°C. It was observed that the month of October had the highest absolute maximum temperatures compared to the other two months. In November, room B had the highest absolute maximum temperature (32.7°C), followed by rooms A (32.6°C) and C (31.4°C). In December, the room with the highest absolute maximum temperature was room B with 33.3°C, followed by room C with 32.6°C and A with 32.1°C.

The months with the highest absolute maximum temperatures were October, December and November respectively. October is a period when rainfall tends to increase in the region, but temperatures don't suffer as much of an impact, since October tends to receive less rainfall than the other months (November and December), precisely because it is the period when the drought period ends and the rainy season begins (Graph 2).

Graph 2 - Absolute maximum air temperatures from October to December 2015

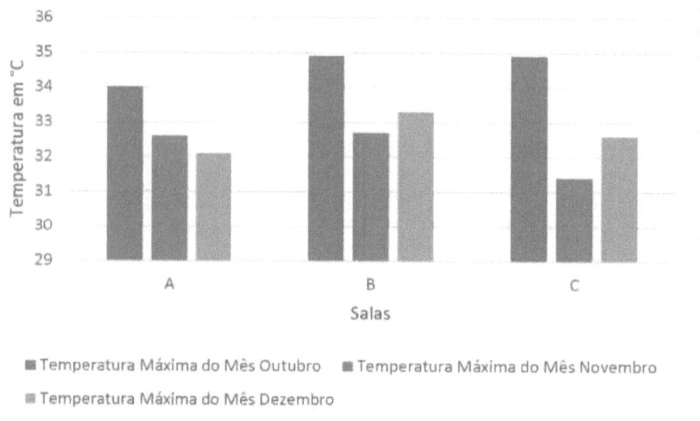

Author: FELTRIN, 2016

The lowest absolute minimum temperature in October was in room A (25.9°C), B (26.8°C) and C (27.6°C). In November, room A again had the lowest absolute minimum temperature (24.3°C), followed by B (25.0°C) and C (26.3°C). In December, the room with the lowest absolute minimum temperature was A (24.6°C), B (25.1°C) and C (25.5°C) simultaneously. This means that October was the month with the lowest absolute minimum temperatures during the study period and that room A was the room with the lowest absolute minimum temperatures compared to the other rooms (Graph 3).

Graph 3 - Absolute maximum air temperatures from October to December 2015

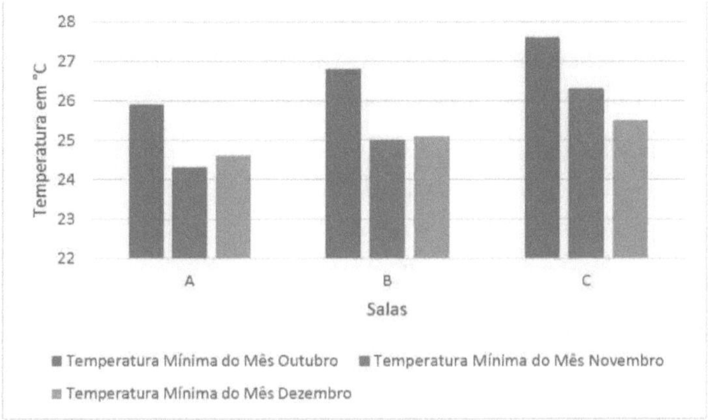

Author: FELTRIN, 2016

2.8.2 Absolute maximum and minimum relative humidity

The absolute maximum relative humidity in October was 75% in room A, followed by rooms C (73.6%) and B (72.4%). In November, the highest absolute relative humidity was in room C (78.5%), followed by rooms A (77.7%) and B (77.5%). In December, the room with the highest absolute relative humidity was in room

A (82.3%), followed by room C (81.3%) and B (80.6%). December is considered to be the month with the highest absolute relative humidity (Graph 4).

Graph 4 - Maximum absolute relative humidity in rooms B, A and C from October to December 2015.

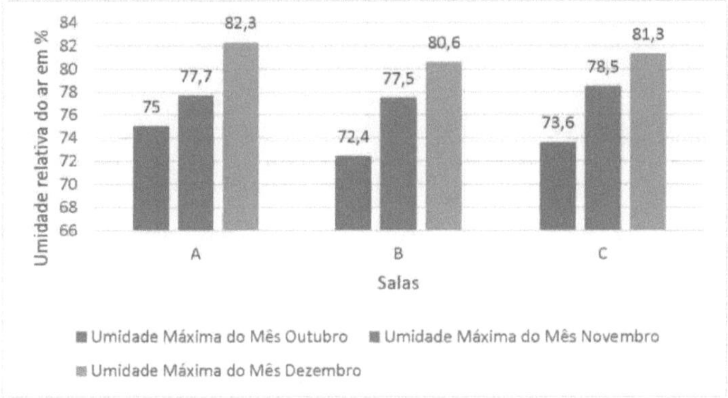

Author: FELTRIN, 2016

The lowest absolute minimum relative humidity in October was in room C (28.2%), followed by rooms A (31.1%) and B (38.4%). In November, the lowest absolute relative humidity was in room

B (44.7%), followed by rooms A (46.9%) and C (49.5%). In December, the room with the absolute lowest relative humidity was in room A (48.1%), followed by rooms B (48.9%) and C (51.1%). October is considered to be the month with the lowest absolute relative humidity values.

During this period it is normal for the humidity to be lower, as the region is going through a period of transition from drought to rain, but it is understood that part of the humidity in the room may be sweat from the approximately 38 students in each room (Graph 5).

Graph 5 - Absolute relative humidity minima for rooms B, A and C from October to December 2015.

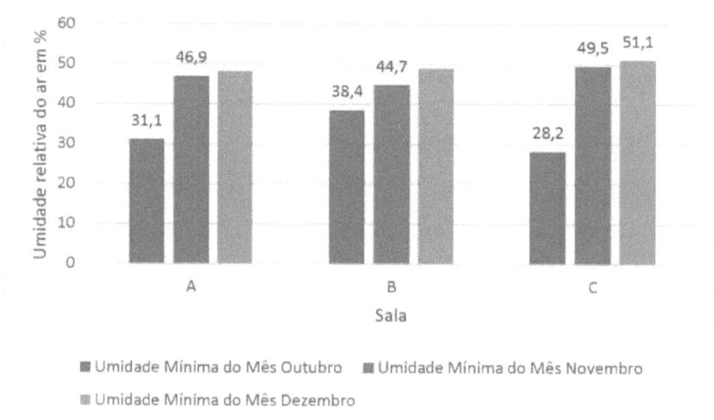

Author: FELTRIN, 2016

When the humidity is low inside the classroom, students can have respiratory problems such as difficulty breathing or even nosebleeds, so humidity is an important factor for thermal comfort. However, the relative humidity did not reach health risk levels in any of the classrooms, according to CEPAGRI (2015).

3.3.3 Dew point absolute highs and lows

The highest absolute dew point in October was in room C (23.6°C), followed by rooms A (23.4°C) and B (23.3°C). In November, the highest absolute dew point value was in room B (25.3°C), followed by rooms C (25.2°C) and A (24.9°C). In December, the highest absolute dew point was in room C (25.8°C), followed by rooms B (25.2°C) and C (23.6°C) (Graph 6).

Graph 6 - Absolute dew points in rooms B, A and C from October to December 2015.

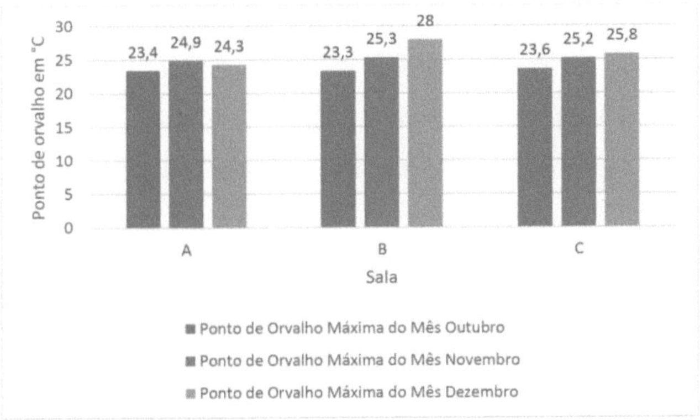

Author: FELTRIN, 2016

The lowest absolute dew point value in October was in room C at 11.8°C, followed by room A at 13.9°C and B at 18.2°C.

In November, the lowest absolute dew point value was in room A (15.8°C), followed by rooms C (16.8°C) and B (17.3°C).

In December, the lowest absolute dew point occurred in room A (14.9°C), followed by rooms B (15.0°C) and C (18.3°C) (Graph 7).

Graph 7 - Absolute minimum dew points in rooms B, A and C from October to December 2015.

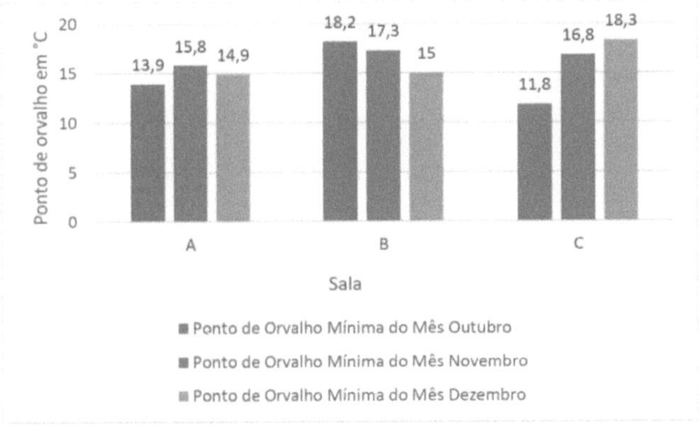

Author: FELTRIN, 2016

3.4 Thermal sensation

For the classroom study, the clothing worn by the students (uniform) was considered; jeans and a short T-shirt, underwear, socks and closed shoes. The variant of 0.761clo was adopted for the thermal resistance of the clothing; activity is sitting and writing, adopting a metabolic rate of

58w/m^2 ; dry bulb temperature (Tbs)[2] in °C, air speed (m/s), to find the PMV and PPD values.

However, the Analysis 1.5 program developed by the Laboratory of Energy Efficiency in Buildings (LABEEE) at the Federal University of Santa Catarina was used for these calculations.

Using the results obtained, the average temperature in each room was calculated and PMV and PPD were calculated. The rooms with the highest PPD in October were B and A, with 66.2%, while C had 59.5%. In November, room B had a PPD of 50.9%, followed by B with 46.2% and C with 37.2%. In December, the highest PPD occurred in B with 51.5%, A with 40.5% and C with 39.6% respectively (Table 1 and Chart 3).

Table 1 - PMV and PPD indices in the rooms

	October		November		December	
Rooms	PMV	PPD	PMV	PPD	PMV	PPD
A	+ 1,6	66,2%	+ 1,4	46,2%	+ 1,3	40,5%
B	+ 1,7	66,2%	+ 1,5	50,9%	+ 1,5	51,5%
C	+ 1,6	59,5%	+ 1,2	37,2%	+ 1,2	39,6%

Author: FELTRIN, 2016.

In practice, PMV is the index that reveals whether the individual is experiencing thermal discomfort from heat or cold and PPD is the number of individuals expressed as a percentage who are feeling dissatisfied in that environment. It should be remembered that according to ASHARAE, only an environment where ≥80% of the individuals feel thermally satisfied is considered comfortable and Table 1 shows that in all months the PMV was positive, so the individuals were feeling discomfort from the heat.

3.5 Students' perceptions

Questionnaires were administered to students in classrooms C, A and B, asking about thermal comfort. In classroom C, 38 students were present on the day the questionnaire was administered, resulting in 95% of the students enrolled in the class. Classroom A had 36 students present with 90% of the students enrolled and classroom B had 37 students representing 93% of the students enrolled.

Analyzing the clothes that the students consider most comfortable, it was found that for classes B and C the short T-shirt was the most popular, while for students in class A it was Bermuda shorts.

---

2 According to Grandi (2006, p. 27), dry bulb temperature (dbT), expressed in degrees Celsius (°C), is the temperature of the air around the human body. Measuring this temperature is important when analyzing heat exchange by convection between the environment and a person's body1.

Among the clothes highlighted in this study, the skirt was the least popular among all the members of the classes, but it should be borne in mind that this is an exclusively female garment (Graph 8).

Graph 8 - Which clothes do you find most comfortable to wear in the classroom according to thermal comfort?

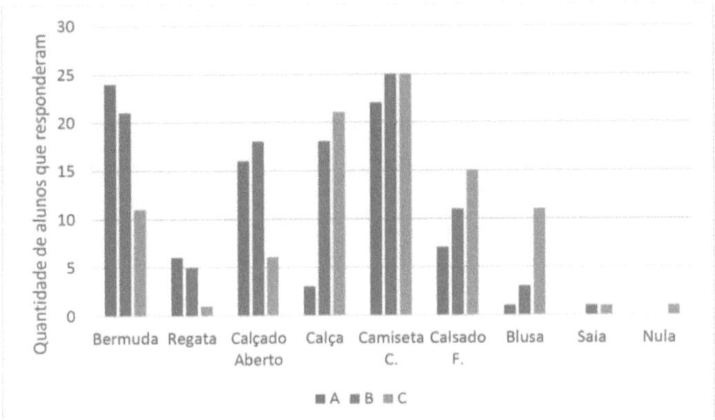

Author: FELTRIN, 2016

The three classes selected for this study consider the room to be warm most of the time. Class C had a high percentage of students who considered the room to be comfortable, as shown in Graph 9. This class has a slab ceiling and windows painted gray, which is a particular feature of the room that may have contributed to the number of students who found it pleasant.

Graph 9 - How do you feel about the temperature in the room most of the time?

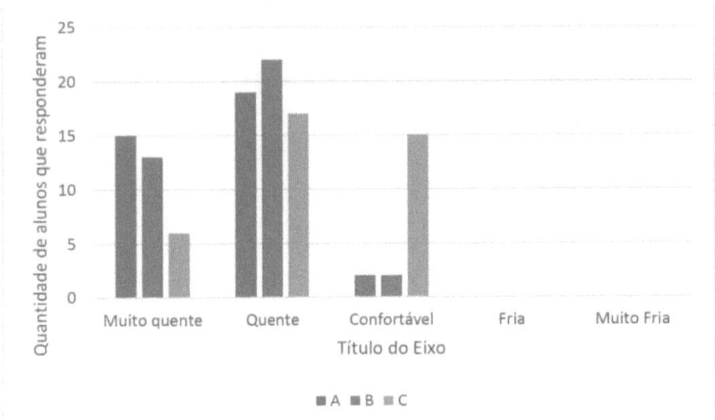

Author: FELTRIN, 2016

For classes B, A and C, natural ventilation (which comes in through the windows) is considered good by the majority of students (Graph 10). However, it can again be seen that class C has a high number of students who consider it to be fair. Looking at the outside area of the room, we notice that there is an access ramp to the upper floor which is above the windows, casting shade and possibly hindering air circulation (Photo 6). In this way, natural ventilation is essential for better thermal comfort in the classroom.

Graph 10 - You like the natural ventilation (coming in through the windows) in the room:

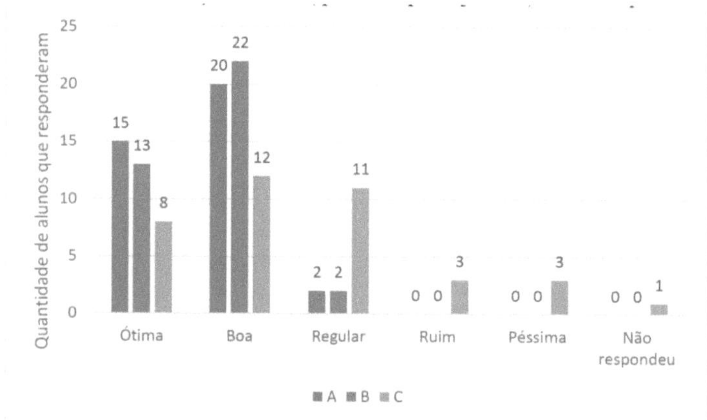

Author: FELTRIN, 2016

Photo 6 - Upper access ramp on the right and the classroom on the left with the windows painted gray for class C

Author: FELTRIN, November 10, 2015.

In the view of students in classes B, A and C, the use of fans in the classroom does not cause discomfort (Graph 11).

Graph 11 - Does the use of fans cause you any discomfort?

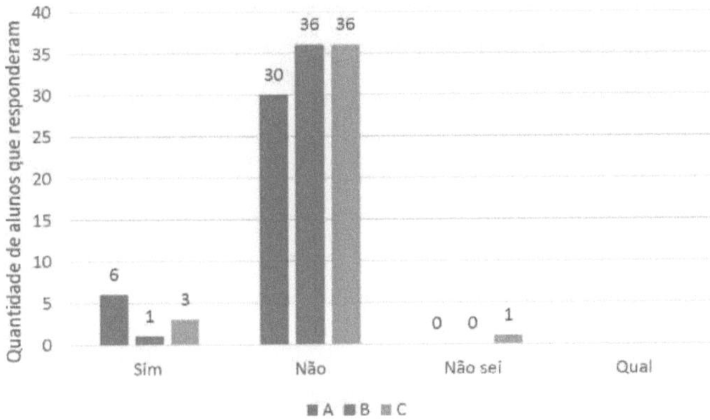

Author: FELTRIN, 2016

In class B, the students considered the use of fans to be good for better thermal comfort. In class A, most of the students answered that the use of the fans was good or fair, and in class C, the students said that the main effect of the fans was fair in terms of thermal comfort. In the case of C, as *already* mentioned, there are a number of students who consider the classroom to be pleasant, so they may have considered the use of fans to be regular since they consider the thermal sensation to be pleasant. It can also be seen that class A had the highest number of bad or poor ratings when taken together (Graph 12).

Graph 12- The effect of fans on thermal comfort in the classroom, in your opinion, is:

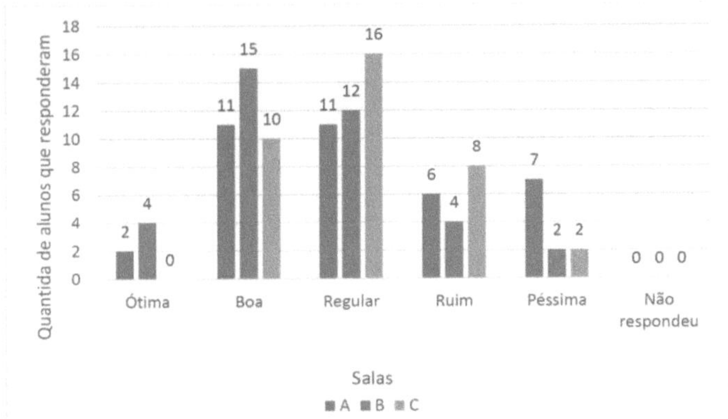

Author: FELTRIN, 2016

The natural lighting in B's classroom is seen by the students as fair. In class A, it is good and in class C the students consider the natural lighting in the classroom to be excellent and good. It should be noted that class C has painted windows and even so, in the opinion of the students, the lighting is good or excellent (Graph 13).

Graph 13 - How would you rate the natural lighting in this room?

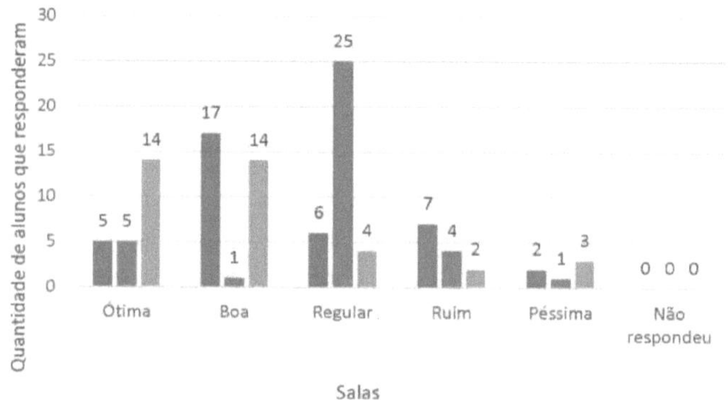

Author: FELTRIN, 2016

The largest number of students from B, A and C consider the number of windows and the way they are distributed in the classroom to be good (Graph 14).

Graph 14 - How would you rate the number and distribution of windows in this room?

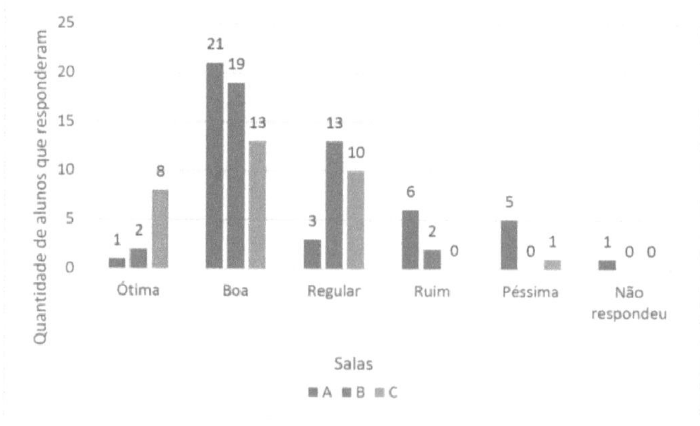

Author: FELTRIN, 2016

Most of the students in B and A classify the artificial lighting in the classroom as good and those in C as excellent, as shown in Graph 15. One factor that could interfere with C's response is the fact that the windows are tinted and so they have to use artificial lighting more than the other classrooms.

Graph 15 - How would you rate the artificial lighting in this room?

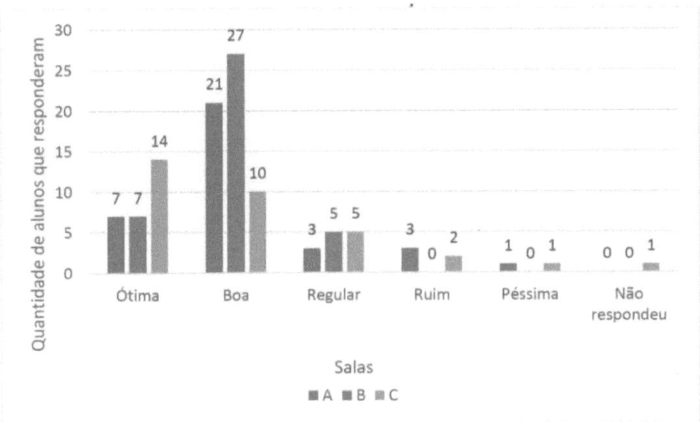

Author: FELTRIN, 2016

According to Graph 16, the majority of students in A and C consider that it is always necessary to use artificial lighting in the classroom, while B considers it to be almost always necessary. The answers from classroom C can be attributed to the fact that the classroom is darker because the window panes are painted, preventing light from entering.

Graph 16 - How often do you think it is necessary to use artificial lighting in this room?

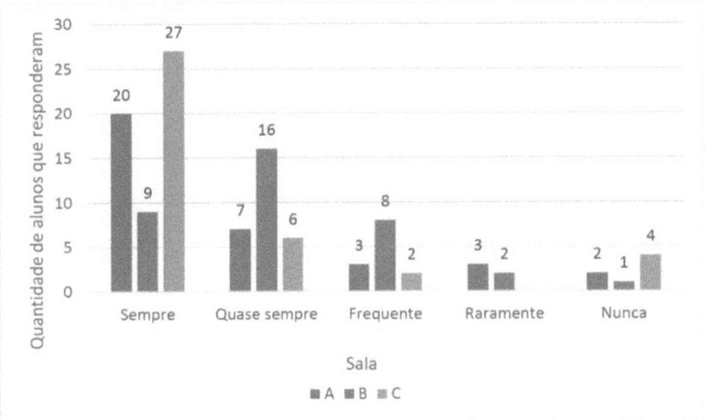

Author: FELTRIN, 2016

With regard to the frequency with which the students consider it necessary to use fans, most of the students in classrooms B, A and C answered that it should always be done, and only one student in class C believed that it should never be done (Graph 17). The students' opinion is explained by the positive PMV indicating thermal discomfort, as shown in Table 1.

Graph 17 - How often do you think it is necessary to use the fan in this room?

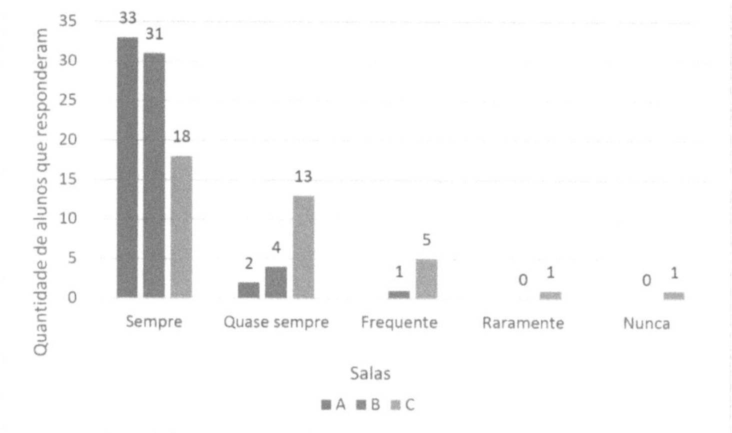

Author: FELTRIN, 2016

Another issue pointed out by the students is that in classrooms B and A, 20 students can easily carry out activities without artificial light on, and 10 students cannot. In room C, the largest number of students said that they could not easily carry out activities without artificial light on, which is contradictory since most of them considered the natural lighting to be excellent or good (Graph 18 and Graph 15).

Graph 18 - Can you easily carry out all the activities without the artificial light on?

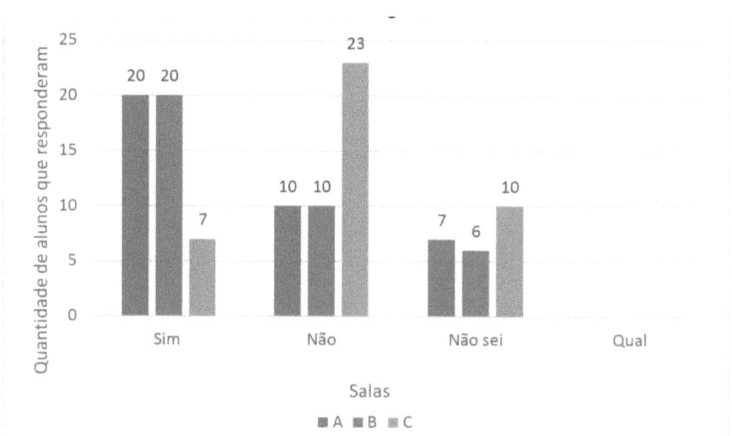

Author: FELTRIN, 2016

When students from B, A and C were asked which part of the room is the hottest, the main answer among the three classes was the back. In this question, 20 students from class A cancelled the answer by answering more than one option to the question: for them, the front, middle and back are hot, as can be seen in Graph 19.

Graph 19 - Which part of the room do you find the hottest?

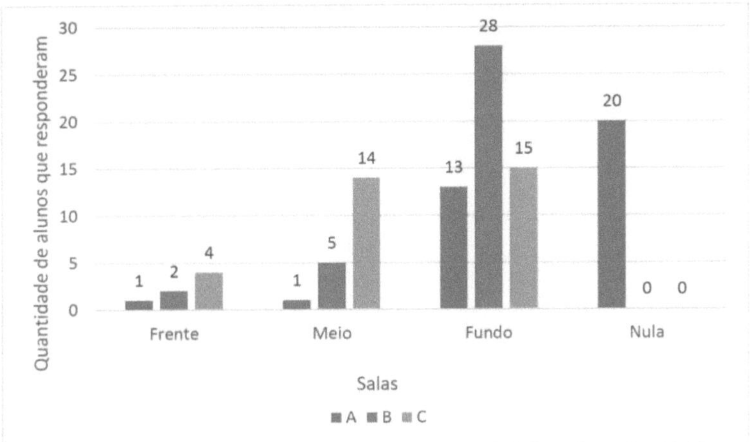

Author: FELTRIN, 2016

When analyzing the students' perception of which of the types of ceiling tiles present in the rooms are best for thermal comfort, the majority of students in classes B, A and C stated that the slab is the best ceiling tile. It should be remembered that the slab ceiling tile is suggested by NBR 15220-3 (2003) (Graph 20).

Graph 20 - The school has two types of ceilings, slab and wooden, which one do you consider the best for thermal comfort?

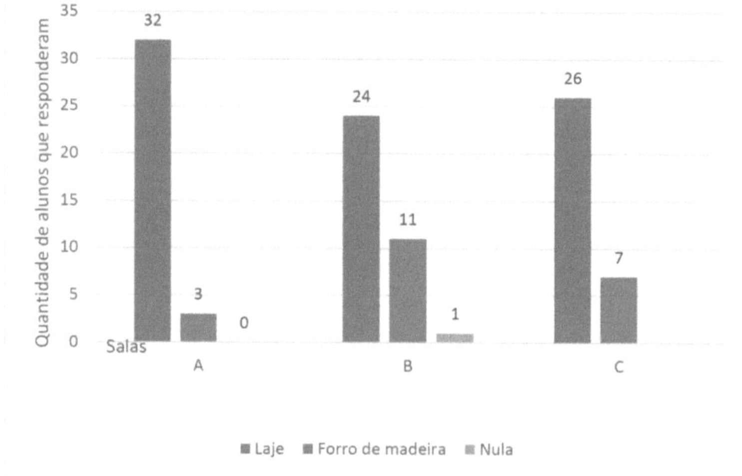

Author: FELTRIN, 2016

It should be noted that the question shown in Graph 21 was open-ended, where students wrote down where they thought it was hottest in the school. Even though the question was not objective, most of the students in classes A and B considered the classroom to be the hottest. Thus, one of the factors for the students naming the classroom as the hottest place in the school is related to the fact that they spend more time in the classroom and because of the greater number of people in the same environment, while for class C, the school yard was considered the hottest place, given that, for a large number of students, the classroom is considered comfortable, as already mentioned (Graph 9).

Graph 21- Which part of the school do you find the hottest?

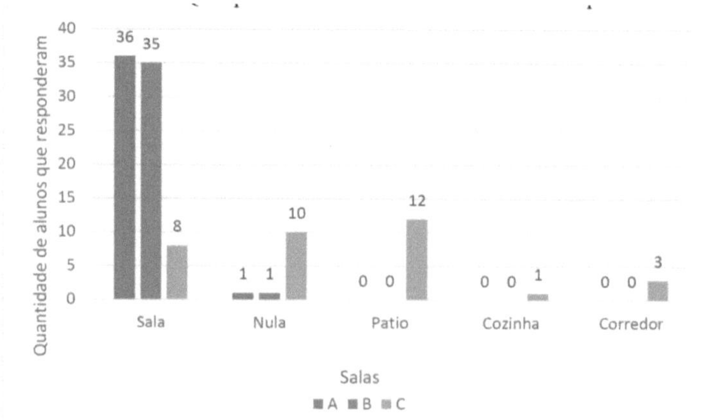

Author: FELTRIN, 2016

Two other questions were only carried out in room C, because only there were the conditions to answer them, only they were exposed to those conditions.

For question 15 of the questionnaire (Appendices B) "Do you think that the painting of the glass on the windows has harmed the lighting in this room?", the largest number of students answered no, this was confirmed in Graph 22, but the number who answered yes is almost the same, which may answer the contradiction between Graphs 13 and 16, where the students first answered that the natural lighting was good or excellent (Graph 13), and later the majority answered that they were unable to carry out activities without artificial light (Graph 16).

In question 16 of the questionnaire (Appendices B), "Do you think that painting the glass on the windows interfered with the temperature in the classroom?", the majority answered that painting the glass should not have interfered with the temperature in the classroom. It should be noted that the students had never been exposed to the classroom without the painted glass, so they had to create a hypothetical scenario of what the classroom would be like without the painted glass and what its temperature would be like, making the answer to this question a little more difficult to elaborate (Graph 22).

Graph 22 - Questions 15 and 16 - Do you think that painting the glass in the windows has affected the lighting in the classroom? 16 - Do you think that painting the glass in the windows has affected the

temperature in the classroom?

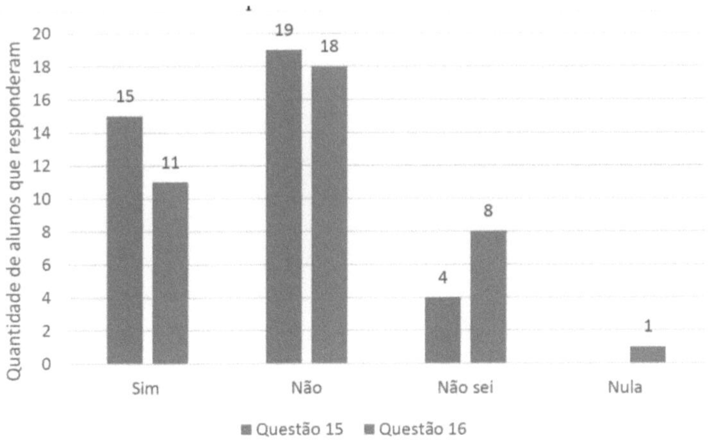

Author: FELTRIN, 2016

By analyzing all the tables and figures, we can see that the majority of students agree that the classroom could have better thermal conditions if more fans were installed; that if the ceilings in all the classrooms were made of slabs, the temperatures would be lower and that all this could lead to better academic performance by the students, since most of them would be feeling thermally comfortable.

4. FINAL CONSIDERATIONS

From all the data presented, we can see that:

- All PPDs were above (for thermal discomfort due to heat) those recommended by ASHRAE;
- In classrooms A and B, the majority of students find it very hot or hot.
- There are not enough fans in all the rooms, so increasing the number of fans could alleviate the discomfort.
- The layout and number of windows were considered by the students to be good or fair, and they meet the recommendations of NBR 15220-3.
- Students consider fans indispensable in the classroom to alleviate thermal discomfort.
- The slab ceiling is better than the wooden ceiling in the opinion of the students and this may have been the main factor in room C having less thermal discomfort than the other rooms.
- Most students consider the back of the room to be the hottest place, so installing fans there could reduce discomfort.
- Classrooms A and B consider the classroom to be the hottest place in the whole school, reinforcing their dissatisfaction.
- All the walls of the school block do not meet the specifications of NBR 15220-3, possibly to save public money on building materials.
- The rooms with slab ceilings do not have high openings for the chimney effect to occur and help with the exchange of loads and gases.
- The average MPV was 1.4% and the average PPD was 50.9%, demonstrating thermal discomfort for more than half of the classroom.
- One possible solution to the thermal discomfort would be to install air-conditioning units to reduce the temperature when necessary and thus provide better thermal comfort and possibly better teaching performance for teachers and students.

REFERENCES

AYOADE, J. **Introduction to climatology for the tropics**. 9ª ed. Rio de Janeiro: Bertrand Brasil, 2003.

BARBOSA. H. P., AMORIM M. C. de C. T. Clima urbano em Presidente Prudente/SP: diferenças térmicas e higrométricas urbano/rural em episódios de outono. **Revista Geonorte**, Special Issue 2, V.2, N.5, p. 220 - 232, 2012

BECERRA, J. A. B.; SHIMABUKURO, Y. E.; ALVALA, R. C. DOS S. Relationship between seasonal vegetation pattern and precipitation in the cerrado region of the Legal Amazon, using vegetation spectral indices. **Revista Brasileira de Meteorologia**, v.24, n.2, 125-134, 2009.

CHRISTOFOLETTI, A. L. H.; HORTA, I. T. L. G.; ANDRÉ, I. R. N. Analysis of thermal comfort indices applied in classrooms: A case study at the Professor Marciano de Toledo Piza State School, in the municipality of Rio Claro -SP. **REVISTA GEONORTE**, Special Issue 2, V.2, N.5, p.711 - 721, 2012.

VIANA, Simone Scatolon Menotti. Thermal comfort in state schools in Presidente Prudente/SP. In: **Urban and regional climatology:** theoretical issues and case studies. 1. ed. Sao Paulo: Outras expressoes, p. 221 - 245, 2013.

FALCAO. M. T., PINHEIRO N. M., RODRIGUES R. F. Urban environmental implications of occupation in valley bottoms: a case study in the Igarapé Pricuma watershed in Boa Vista / RR. In. BRAZILIAN SYMPOSIUM OF APPLIED PHYSICAL GEOGRAPHY, 8., 2009, Viçosa. **Proceedings**. Viçosa: Federal University of Viçosa, 2009. 1 CD.

FANGER, P. O. **Thermal comfort: Analysiss and applications in environmental engineering**. United States: McGraw-Hill Book Company, 1970. 244p.

FROTA, Anésia Barros; SCHIFFER, Sueli Ramos. **Manual of thermal comfort:** architecture, urbanism. 5. ed. Sao Paulos: Estudio Novel. 2001.

GRANDI, Mariele Stefani. **Evaluation of thermal perception in a control room**. Porto alegre: UFRG; 2006.

LOMBARDO. M. A. **Ilhas de calor nas metrópoles:** um exemplo de Sao Paulo. Sao Paulo: HUCITEC, 1985.

LOPES, R. M.; MARIANO, Z, de F.; SCOPEL, I. Rainfall variability in the municipality of Jatai-GO (partial result). X BRAZILIAN SYMPOSIUM OF APPLIED PHYSICAL GEOGRAPHY. **Proceedings.** http://www.cibergeo.org/XSBGFA/eixo3/3.4/208/208.htm

MENDONÇA, F. and MONTEIRO, Carlos A. F. **Cima urbano.** Contexto Publishing House, 2003.

MONTEIRO, Carlos Augusto F., **O clima e a organização do espaço no estado de Sao Paulo: problemas e perspectivas.** IGEOG-USP series of theses and monographs N°28, Sao Paulo 1976.

MONTEIRO, C. A. de. F. **Teoria e clima urbano**. 1976. 181 f. Thesis. (Submitted to the competition for the position of Professor in the Department of Geography) - Faculty of Philosophy, Letters and Human Sciences, University of Sao Paulo, Sao Paulo, 1976.

PEREIRA D. A. de M.; LEITE, J. P.; BARBOSA, R. F.; LIMA, O. F. de. Evaluation of the performance of natural ventilation for the promotion of thermal comfort in a public educational block. In: VIII National Congress of Excellence in Management. Rio de Janeiro - RJ, **Anais**... Rio de Janeiro - RJ: 2011.

ROCHA, José Ricardo Rodrigues. Climatic aspects and occurrence of dengue cases in Jatai - GO. **Revista Geonorte**, Special Edition, V.2, N.4, p.757 - 769, 2012.

ROCHA, José Ricardo Rodrigues; MARIANO, Zilda de Fâtima; FELTRIN, Jean Carlos. Climate in small towns: the thermodynamic system of Jatai (GO). **Revista Brasileira de Climatologia**, Special Issue 10 V.15, p. 205 - 228. 2014.

RODRIGUES, Zulimar Màrita Ribeiro. Climate and health implications: analysis of hospitalizations for respiratory diseases in Sao Luis (MA). 2nd Brazilian Symposium on Health & Environment (2ndSIBSA): Belo Horizonte (MG). **Proceedings**; Federal University of Maranhao, 2014.

ROSSATO. P. S. **O sistema termodinâmico do clima urbano de Nova Palma, RS: contribuçâo ao clima urbano de cidades pequenas.** *2010.* 121 f. Dissertation (Master's in Geography) - Federal University of Santa Maria, Santa Maria, 2010.

RUAS, Alvaro C. **Conforto térmico nos ambientes de trabalho**. Brasilia: Fundacentro, 1999.

SILVA, Daniel, Araujo da. The role of precipitation in the comfort of the urban climate of the city of Manaus. **Revista Geonorte**, Special Issue 2, v.2, N.5, p.22 - 40, 2012.

SILVA, J. S, et al. Urban climate of Jatai-Goiàs. In: SIMPÒSIO BRASILEIRO DE CLIMATOLOGIA GEOGRAFICA, 6, 2004. Aracajù-SE. **Proceedings**... Aracajù-SE: UFS, 2004, p.50.

XAVIER, Antônio Augusto de Paula. **Thermal comfort conditions for high school students in the Florianópolis region. 1999.** Dissertation (Post-Graduation in Civil Engineering) Federal University of Santa Catarina, Florianópolis, 1999.

Accessed on December 29, 2015

<http://www.cpa.unicamp.br/artigosespeciais/umidade-do-ar-saude-no-inverno.html>

Accessed December 29, 2015 <http://www.cgesp.org/v3/umidade-relativa-do- ar.jsp>

Accessed December 29, 2015 <http://ciram.epagri.sc.gov.br/index.php?option =comcontent&view=article&id=1498&Itemid=661>

Accessed on January 03, 2016 <http://www.cgesp.org/v3/umidade-relativa-do-ar.jsp>

Accessed February 12, 2016 < http://www.labeee.ufsc.br/downloads/softwares/ Analysiss>

APPENDICES

Appendices A - Questionnaire administered to students

Date of questionnaire: 22/01/2016

Weight: ____kg Age: _______ years Sex: ( ) male ( ) female

Clothing you are wearing: ( ) open shoes ( ) closed shoes ( ) socks ( ) pants ( ) underwear ( ) panties ( ) bra ( ) T-shirt ( ) short-sleeved shirt ( ) long-sleeved shirt ( ) jeans ( ) dress pants ( ) short skirt ( ) long skirt ( ) shorts ( ) thin jacket ( ) thick jacket ( ) jacket ( ) other (please describe):

1 - Which clothes do you find most comfortable to wear in the classroom according to thermal comfort?

____

2 - What is the temperature like in the room most of the time?

Very Warm ( ) Warm ( ) Comfortable ( ) Cold Very Cold ( )

3 - The natural ventilation (coming in through the windows) in the living room is for you:

Great ( ) Good ( ) Fair ( ) Poor ( ) Bad ( )

4 - Does the use of fans cause you any discomfort?

Yes ( ) No ( ) I don't know ( ) Which one? _________________

5 - The effect of fans on thermal comfort in the classroom, in your opinion, is:

Great ( ) Good ( ) Fair ( ) Poor ( ) Bad ( )

6 - How would you rate the natural lighting in this room?

Great ( ) Good ( ) Fair ( ) Poor ( ) Bad ( )

7 - How would you rate the number and distribution of windows in this room?

Great ( ) Good ( ) Fair ( ) Poor ( ) Bad ( )

8 - How would you rate the artificial lighting in this room?

Great ( ) Good ( ) Fair ( ) Poor ( ) Bad ( )

9 - How often do you think it is necessary to use artificial lighting in this room?

10 - How often do you think it's necessary to use the fan in this room?

Always ( ) Almost always ( ) Frequ. Average ( ) Rarely ( ) Never ( )

11 - Can you do all your activities without the artificial light on?

Yes ( ) No ( ) I don't know ( )

12 - Which part of the room do you find warmest?

Front (frame) ( ) Middle ( ) Back ( )

13 - The school has two types of ceilings, slab and wooden, which do you think is the best for thermal comfort?

( ) slab ( ) wooden ceiling

14 - Which part of the school do you think is the hottest?

Appendices B - Questionnaire administered to the students of C

Date of questionnaire: 22/01/2016

Weight: _____ kg Age: _______ years Sex: ( ) male ( ) female

Clothing you are wearing: ( ) open shoes ( ) closed shoes ( ) socks ( ) pants ( ) underwear ( ) panties ( ) bra ( ) T-shirt ( ) short-sleeved shirt ( ) long-sleeved shirt ( ) jeans ( ) dress pants ( ) short skirt ( ) long skirt ( ) shorts ( ) thin jacket ( ) thick jacket ( ) jacket ( ) other (please describe):

1 - Which clothes do you find most comfortable to wear in the classroom according to thermal comfort?

---

2 - What is the temperature like in the room most of the time?

Very Warm ( ) Warm ( ) Comfortable ( ) Cold Very Cold ( )

3 - The natural ventilation (coming in through the windows) in the living room is for you:

Great ( ) Good ( ) Fair ( ) Poor ( ) Bad ( )

4 - Do the fans cause you any discomfort?

Yes ( ) No ( ) I don't know ( ) Which one? __________________

5 - The effect of fans on thermal comfort in the classroom, in your opinion, is:

Great ( ) Good ( ) Fair ( ) Poor ( ) Bad ( )

6 - How would you rate the natural lighting in this room?

Great ( ) Good ( ) Fair ( ) Poor ( ) Bad ( )

7 - How would you rate the number and distribution of windows in this room?

Great ( ) Good ( ) Fair ( ) Poor ( ) Bad ( )

8 - How would you rate the artificial lighting in this room?

Great ( ) Good ( ) Fair ( ) Poor ( ) Bad ( )

9 - How often do you think it is necessary to use artificial lighting in this room?

10 - How often do you think it's necessary to use the fan in this room?

Always ( ) Almost always ( ) Frequ. Average ( ) Rarely ( ) Never ( )

11 - Can you do all your activities easily without the artificial light on?

Yes ( ) No ( ) I don't know ( )

12 - Which part of the room do you find warmest?

Front (frame) ( ) Middle ( ) Back ( )

13 - The school has two types of ceilings, slab and wooden, which do you think is the best for thermal comfort?

( ) slab ( ) wooden ceiling

14 - Which part of the school do you think is the hottest?

15 - (Only for 6th grade) Do you think that the painting of the glass on the windows has affected the lighting in this room?

Yes ( ) No ( ) I don't know ( )

16 - (Only for 6th grade) Do you think that painting the glass on the windows has affected the temperature in the classroom?

Yes ( ) No ( ) I don't know ( )

Photo 7- School block on the south side (A, B and D); Window in room B (C); Corridor between rooms (E); Type of windows (F); Rooms with wooden ceilings (G); Rooms with slab ceilings (H) Fiber cement roof (I)

Author: FELTRIN, November 10, 2015.

ANNEX

Annex I - Metabolic rate for different activities

Metabolisms for different activities		
Activity	**Metabolic rate**	
Office activity		
Reading, sitting	55 W/m^2	2 1.0 W/m2
Writing	60 W/m2	2 1.0 W/m2
Typing	65 W/m2	2 1.1 W/m2
Filing, sitting	70 W/m2	2 1.2 W/m2
Filing, standing	80 W/m2	2 1.4 W/m2
Walking	100 W/m2	2 0.7 W/m2
Resting		
Sleeping	40 W/m2	0.7 MET
Reclining	45 W/m2	0.8 MET
Sitting still	60 W/m2	2 1,0 MET
Standing, relaxed	70 W/m2	2 1,2 MET
Driving/flying		
Car	60 to 115 W/m2	2 1.0 to 2.0 MET
Aviation, routine	70 W/m2	2 1,2 MET
Aircraft, landing	105 W/m2	2 1,8 MET
Aviation, combat	140 W/m2	2.4 MET
Heavy vehicle	185 W/m2	2 3,2 MET

Source: ASHRAE, 2005

Organizer: FELTRIN, 2016.

MIX
Papier aus verantwortungsvollen Quellen
Paper from responsible sources
FSC® C105338

Printed by Books on Demand GmbH, Norderstedt / Germany